How To Get into Optometry School

By Shawn Matsumoto

www.preopto.com

Disclaimer
I have changed the names to protect individuals' privacy.
To maintain the anonymity of individuals, I have omitted
or changed some details. This book does not guarantee
admission to any optometry schools. The book does not
replace the advice of any college counselor, admissions
advisor, or professional advisor. The information of this
book was correct at the time of publication, but the author
does not assume any liability for loss or damage caused by
errors or omissions.

Table of Contents

Chapter I: What is Optometry School?

Before we dive deeper into what optometry school is, I think it's essential for us to talk about the practice of optometry and what it means to be an optometrist. Optometry is more than just a medical profession, and it is more than just being a doctor – it is about providing comfort to the lives of people who rely heavily on you. As cliché as that sounds, the more you think about it, the more it becomes apparent. While patients may go to an orthopedist for gradual healing of a fractured bone, patients usually go to optometrists for immediate changes in their quality of life. While ophthalmologists often see patients for only five minutes, optometrists would often see patients for fifteen, twenty minutes at a time. While there may not be a significant foundation of a doctor-patient relationship in urgent care, patients often bond with the doctor in optometric settings.

In the eyes of an optometrist, the patient is not just another number or another case. They are real people. The quality of interactions required in optometry allows the doctor to see the eyes of the patient and the entire person – their mental state, everyday lives, passions, and aspirations. The connection between the doctor and patient creates an open and comforting environment not available in other physicians' offices. The bond made here is unique and personal that visiting the optometrist is something to look forward to rather than dread.

Yet, optometry is a medical profession. It's not simply about writing a glasses prescription and calling it a day – it's about understanding why the patients are seeing a certain way, keeping in mind the possibilities of other pathologies that may initially seem unrelated to vision. When they say the eye is the gateway to the body, it is 100% true. Many systemic pathologies can be found first and monitored in the eye, so spotting these minor changes could be crucial to the patients' lives.

Now to the question of "What is Optometry School?": Optometry school is medical school, and optometry school is rigorous. Sure, skeptics might say it's easier to get in or that it's lazy ophthalmology, but none of those people have ever been to optometry school and have never been an optometrist. It's a rigorous, four-year program dedicated to producing doctors capable of treating patients with the knowledge and respect they deserve.

Optometry school is a full-time job, and with all due respect to university institutions, undergraduate school is a joke. Procrastination, sacrificing an exam for the sake of another, and skipping classes simply because you're sleepy is not an option. Expecting to have a sorority-like social life is impossible. Your grades and your capabilities as a doctor will reflect every action you take during optometry school.

Optometry school is full of "expectations." Expectations from professors, your future peers, your school, your patients, and yourself. Once you get into a school, they already have a set of expectations for you that you should acknowledge. They chose you for a reason – and it's not for you to be stagnant and show no growth during your four years there. They chose you because they believed you could succeed in a rigorous program, and they will hold you up to those standards. Your school will expect you to maintain professionalism the moment you arrive on campus. You are now one of their members, their coworkers, their peers. You are now the representative of the school you attend, and they expect you to act in such a manner as not to taint the school's reputation.

Patients have certain expectations for you, especially once you are in rotations and have enough knowledge to tend to the patients yourself. Sure, being lost as a first-year student is unavoidable – you know absolutely nothing about asteroid hyalosis or how to work a slit lamp. Even then, you need to be honest and professional while actively observing and learning how each doctor performs each procedure so that you are more than capable of examining your patients one day.

Yet, the most important expectations are the ones you set for yourself. Setting it too high can be dangerous to your health, while setting it too low will affect your learning negatively. You must find a good balance of what you expect for yourself without expecting perfection.

Even though the program is demanding, optometry school is exciting. You finally learn about what you're passionate about, and you become dedicated to your studies more than ever. You have classmates who become your best friends. You're in a new environment with new experiences, and you get to learn skills that will make you realize that you're almost a Doctor of Optometry.

Optometry school is where you can see the light at the end of the tunnel. The knowledge and skills you learn finally come into play, and you realize that schooling is almost over. Forever. Soon, you get to be the one asking patients which lens is better: one or two.

Chapter II: Do I want to go to Optometry School?

You're most likely an undergraduate student, a recent graduate, or someone who has been in the field of optometry for quite some time, and you have probably thought enough about going into optometry for you to buy this book. After days of contemplating, you've realized why you wanted to be an optometrist. Every one of you probably has your reasons for wanting to become one. You might be doing it for the money, or you might be doing it simply because you saw "optometrist" as one of the most satisfying jobs in the United States. You might not even like eyes, yet you chose to try out the application process and go into this field. For those of you who are passionate about optometry, feel free to skip this chapter. For those still thinking about it, stick around, and let me ask you a few questions to make sure you are willing to employ enthusiasm and commitment to becoming the best doctor you can be.

I'm not here to tell you the facts about optometry and the ins and outs of being an optometrist. There are honestly too many variations of practice that it would be difficult to simplify into this small, "how-to" book. Optometry school will present all the unique career paths anyway, and you will have four full years to experience and learn about each type of practice. I am also not here to tell you why you should choose to become an optometrist and why you shouldn't, because, well, I'm going to be brutally biased. What I am here to do, though, is to jog your mind and make you consider your commitment to the field.

1. Will you be proud to call yourself a Doctor of Optometry?

Being motivated and having pride is vital in treating patients in any mode of practice. It's not a part-time server job where you can complain about large tables and simply quit because you're "tired" of customers. Bad days will happen, and you must stay motivated and connected with each patient for the rest of your professional career. Let's say you go through optometry school, get your degree, find a job, start prescribing eyeglass prescriptions, and after your first day, you feel no pride and joy. You're embarrassed to name yourself an OD, and you dread talking to patients. You start complaining about each patient, and you continue to pass your patients on to another doctor just because you "don't want to deal with people right now." Even after four years of hard work and dedication, you end up unmotivated and disconnected from the practice. So, now the question is, where did your four years and thousands of dollars go? Before you make that mistake, I want you to think about how excited you are to be on the path of becoming a certified OD and how proud you would be to serve patients for the rest of your career.

2. Are you willing to invest both time and money in yourself?

Optometry school isn't free, full-ride scholarships are unavailable, and early graduation is not an option. You're stuck in the school, with your few classmates, learning something new every day for the next four years. You already went through four years of college. You already have a degree. You may already have student loans, and you don't want to learn about DNA replication again. Well, ready for another round?

3. Are you willing to actively learn and keep an open mind?

Open-mindedness is important, especially if you are working with patients. Every individual you interact with – doctors, colleagues, or patients, has different backgrounds, cultures, and experiences. Being their doctor, you must be welcoming to all and must treat each patient respectfully and professionally. Misdiagnosing patients due to their background or providing inadequate care based on their appearance is morally and politically unacceptable. Being open and learning about the different people in this world will be valuable to your future as an optometrist.

Now, I didn't bring all this up to scare you away. It might seem like this field is asking for too much, but these are essentially all requirements for you to enter the professional world. Your name as a doctor will become more than just a name. Once you enroll in an optometry school, you would represent yourself, your colleagues, your practice, and your school, so having certain expectations and requirements will be necessary to become a reliable optometrist.

Chapter III: Optometry Schools

Optometry schools

There are currently 23 total accredited universities providing optometric education. Remember that *each school has different requirements*. Please check OptomCAS and the school website to confirm their requirements for courses, letters of recommendation, and supplemental applications. These requirements are subject to change every year. The data presented below are from the 2021 applicant class, and the NBEO pass rate is from the class of 2020. IUSO did not provide data for the matriculating class of 2021 due to changes in admission requirements due to Covid-19. *This information is from the ASCO Profiles of Applicants for 2021.*

Alabama

University of Alabama at Birmingham School of Optometry (UABSO)
Location: Birmingham, Alabama
Average GPA: 3.69
Average OAT TS: 321
Average OAT AA: 328
Other tests that are accepted: MCAT, DAT, PCAT
Bachelor's Degree Required?: Preferred
Regional Tuition: $28,915
Non-Regional Tuition: $55,069 (no scholarship), $40,071 (merit scholarship)
Applied/Admitted: 420/121
NBEO Pass Rate: 97.87%

Arizona

Arizona College of Optometry, Midwestern University (AZCOPT)
Location: Glendale, Arizona
Average GPA: 3.54
Average OAT TS: 320
Average OAT AA: 326
Other tests that are accepted: MCAT, DAT, PCAT, GRE
Bachelor's Degree Required?: Yes
Regional Tuition: $49,676
Non-Regional Tuition: $49,676
Applied/Admitted: 658/218
NBEO Pass Rate: 98.00%

California

Southern California College of Optometry at Marshall B Ketchum University (SCCO)
Location: Fullerton, California

Average GPA: 3.52
Average OAT TS: 348
Average OAT AA: 348
Other tests that are accepted: MCAT, DAT, PCAT, GRE
Bachelor's Degree Required?: No
Regional Tuition: $47,025
Non-Regional Tuition: $47,025
Applied/Admitted: 546/214
NBEO Pass Rate: 91.84%

University of California – Berkeley School of Optometry (UCBSO)
    Location: Berkeley, California
    Average GPA: 3.59
    Average OAT TS: 348
    Average OAT AA: 348
    Other tests that are accepted: MCAT, DAT, PCAT, GRE
    Bachelor's Degree Required?: Yes
    Regional Tuition: $41,472
    Non-Regional Tuition: $53,493
    Applied/Admitted: 305/88
    NBEO Pass Rate: 92.54%

Western University of Health Sciences College of Optometry (WesternU)
    Location: Pomona, California
    Average GPA: 3.27
    Average OAT TS: 313
    Average OAT AA: 318
    Other tests that are accepted: MCAT, DAT, PCAT, GRE
    Bachelor's Degree Required?: No
    Regional Tuition: $46,159
    Non-Regional Tuition: $46,159
    Applied/Admitted: 395/141
    NBEO Pass Rate: 88.73%

Florida

NOVA Southeastern University College of Optometry (NOVA or NSUCO)
    Location: Fort Lauderdale, Florida
    Average GPA: 3.48
    Average OAT TS: 325
    Average OAT AA: 321
    Other tests that are accepted: None
    Bachelor's Degree Required?: No
    Regional Tuition: $36,770
    Non-Regional Tuition: $41,343
    Applied/Admitted: 828/192
    NBEO Pass Rate: 97.62%

Illinois

Chicago College of Optometry, Midwestern University (CCO)
    Location: Downers Grove, Illinois
    Average GPA: 3.42
    Average OAT TS: 301
    Average OAT AA: 310
    Other tests that are accepted: MCAT, DAT, PCAT, GRE
    Bachelor's Degree Required?: Yes
    Regional Tuition: $45,188
    Non-Regional Tuition: $45,188
    Applied/Admitted: 600/259
    NBEO Pass Rate: NA

Illinois College of Optometry (ICO)
    Location: Chicago, Illinois
    Average GPA: 3.36
    Average OAT TS: 317
    Average OAT AA: 323

Other tests that are accepted: GRE
Bachelor's Degree Required?: Strongly Recommended
Regional Tuition: $44,336
Non-Regional Tuition: $44,336
Applied/Admitted: 1008/347
NBEO Pass Rate: 92.54%

Indiana

Indiana University School of Optometry (IUSO)
    Location: Bloomington, Indiana
    Average GPA: 3.59
    Average OAT TS: 319
    Average OAT AA: 327
    Other tests that are accepted: GRE, MCAT
    Bachelor's Degree Required?: No
    Regional Tuition: $31,838
    Non-Regional Tuition: $44,374
    Applied/Admitted: 752/195
    NBEO Pass Rate: 89.80%

Kentucky

University of Pikeville – Kentucky College of Optometry (KYCO)
    Location: Pikeville, Kentucky
    Average GPA: 3.44
    Average OAT TS: 300
    Average OAT AA: 312
    Other tests that are accepted: MCAT, DAT, PCAT
    Bachelor's Degree Required?: No
    Regional Tuition: $44,000
    Non-Regional Tuition: $44,000
    Applied/Admitted: 483/236
    NBEO Pass Rate: 91.89%

Massachusetts

Massachusetts College of Pharmacy and Health Sciences School of Optometry (MCPHS)
Location: Worcester, Massachusetts
Average GPA: 3.40
Average OAT TS: 311
Average OAT AA: 317
Other tests that are accepted: GRE
Bachelor's Degree Required?: No
Regional Tuition: $45,360
Non-Regional Tuition: $45,360
Applied/Admitted: 550/266
NBEO Pass Rate: 83.67%

New England College of Optometry (NECO)
Location: Boston, Massachusetts
Average GPA: 3.52
Average OAT TS: 332
Average OAT AA: 335
Other tests that are accepted: GRE, MCAT, DAT
Bachelor's Degree Required?: No
Regional Tuition: $44,119
Non-Regional Tuition: $44,119
Applied/Admitted: 842/404
NBEO Pass Rate: 88.03%

Michigan

Michigan College of Optometry at Ferris State University (MCO)
Location: Big Rapids, Michigan
Average GPA: 3.61
Average OAT TS: 329
Average OAT AA: 334
Other tests that are accepted: None

Bachelor's Degree Required?: Preferred
Regional Tuition: $33,212
Non-Regional Tuition: $33,212
Applied/Admitted: 189/88
NBEO Pass Rate: 90.63%

Missouri

University of Missouri at St. Louis College of Optometry
(UMSL)
  Location: St. Louis, Missouri
  Average GPA: 3.50
  Average OAT TS: 310
  Average OAT AA: 320
  Other tests that are accepted: MCAT, DAT, PCAT, GRE
  Bachelor's Degree Required?: No
  Regional Tuition: $27,100
  Non-Regional Tuition: $43,569
  Applied/Admitted: 374/119
  NBEO Pass Rate: 95.65%

New York

State University of New York College of Optometry
(SUNY)
  Location: New York, New York
  Average GPA: 3.61
  Average OAT TS: 348
  Average OAT AA: 348
  Other tests that are accepted: MCAT, DAT, PCAT, GRE
  Bachelor's Degree Required?: Strongly Preferred
  Regional Tuition: $30,510
  Non-Regional Tuition: $51,840
  Applied/Admitted: 437/166
  NBEO Pass Rate: 96.84%

Ohio

The Ohio State University College of Optometry (OSU)
 Location: Columbus, Ohio
 Average GPA: 3.70
 Average OAT TS: 351
 Average OAT AA: 350
 Other tests that are accepted: MCAT, DAT, PCAT, GRE
 Bachelor's Degree Required?: No
 Regional Tuition: $29,053
 Non-Regional Tuition: $50,109
 Applied/Admitted: 547/92
 NBEO Pass Rate: 96.83%

Oklahoma

Northeastern State University – Oklahoma College of Optometry (NSUOCO)
 Location: Tahlequah, Oklahoma
 Average GPA: 3.62
 Average OAT TS: 304
 Average OAT AA: 316
 Other tests that are accepted: MCAT, DAT
 Bachelor's Degree Required?: No
 Regional Tuition: $21,245
 Non-Regional Tuition: $40,495
 Applied/Admitted: 207/55
 NBEO Pass Rate: 72.00%

Oregon

Pacific University College of Optometry (PUCO)
 Location: Forest Grove, Oregon
 Average GPA: 3.48
 Average OAT TS: 324
 Average OAT AA: 330

Other tests that are accepted: GRE, MCAT, DAT
Bachelor's Degree Required?: No
Regional Tuition: $45,341
Non-Regional Tuition: $45,341
Applied/Admitted: 377/197
NBEO Pass Rate: 87.91%

Pennsylvania

Pennsylvania College of Optometry at Salus University (SALUS or PCO)
Location: Elkins Park, Pennsylvania
Average GPA: 3.41
Average OAT TS: 308
Average OAT AA: 317
Other tests that are accepted: MCAT, DAT, PCAT, GRE
Bachelor's Degree Required?: No
Regional Tuition: $43,415
Non-Regional Tuition: $43,415
Applied/Admitted: 904/439
NBEO Pass Rate: 91.03%

Puerto Rico

Inter American University of Puerto Rico School of Optometry (IAUPR)
Location: Bayamon, Puerto Rico
Average GPA: 3.16
Average OAT TS: 283
Average OAT AA: 297
Other tests that are accepted: GRE
Bachelor's Degree Required?: No
Regional Tuition: $30,140
Non-Regional Tuition: $30,140
Applied/Admitted: 235/123
NBEO Pass Rate: 92.00%

Tennessee

Southern College of Optometry (SCO)
 Location: Memphis, Tennessee
 Average GPA: 3.63
 Average OAT TS: 333
 Average OAT AA: 337
 Other tests that are accepted: GRE, MCAT
 Bachelor's Degree Required?: Strongly Preferred
 Regional Tuition: $20,278
 Non-Regional Tuition: $39,478
 Applied/Admitted: 759/308
 NBEO Pass Rate: 98.50%

Texas

University of Houston College of Optometry (UHCO)
 Location: Houston, Texas
 Average GPA: 3.41
 Average OAT TS: 313
 Average OAT AA: 321
 Other tests that are accepted: GRE, DAT, MCAT
 Bachelor's Degree Required?: No
 Regional Tuition: $40,385
 Non-Regional Tuition: $40,385
 Applied/Admitted: 474/195
 NBEO Pass Rate: 90.11%

University of the Incarnate Word, Rosenberg School of Optometry (UIWRSO)
 Location: San Antonio, Texas
 Average GPA: 3.63
 Average OAT TS: 337
 Average OAT AA: 340
 Other tests that are accepted: MCAT, DAT, PCAT, GRE

Bachelor's Degree Required?: Yes
Regional Tuition: $27,090
Non-Regional Tuition: $46,106
Applied/Admitted: 773/155
NBEO Pass Rate: 88.52%

What to look for in a school

There are no rankings for optometry schools. I repeat. *There are no rankings for optometry schools.*
With that said, go where you want to go, not where you think you should go. Apply to where you are interested, not where people say you should apply. Every school has different tuition, location, class size, teaching methods, course schedule, and clinical experiences. It is your job to think about what is important to you and rank each school based on your preferences.

Let's say you're an applicant who currently lives in Texas with your family, and you value proximity to your home. UHCO and UIWRSO are probably your best bet. Is tuition a crucial factor to you? SCO (for their large scholarship pool) and UHCO (for their in-state tuition) might be schools to keep an eye on. Do you value clinical experience? SUNY and ICO are in the middle of large cities with plenty of different people and cases. Do you want to go into research? Why not give UABSO a chance?

During a college visit, you might vibe with one school more than another, and that's okay. It's entirely up to you to determine which school will give you the most happiness and comfort for the next few years and at which school you think you will learn the greatest. Some online forums might rank the schools with the criteria of the date of accreditation, the total number of graduates, and national board scores. Still, those are just the result of a competitive mindset that humans naturally have. All I have to say about these comparisons is this: the education you receive will only depend on what *you* decide to make out of it, regardless of your school.

Chapter IV: OptomCAS

OptomCAS is the primary means of applying to all accredited optometry schools in the United States (including Puerto Rico). Many schools require supplemental applications, but you would generally complete these after you submit your OptomCAS application. The website contains every detail necessary for you to apply EXCEPT for each schools' information. While the website displays a summary of each school, it is not enough information for you to decide whether to apply to those schools.

Yeah, you're going to be here a lot, submitting paperwork, entering personal information, applying to schools, paying application fees. This website will be your life for the first half of your application cycle, and luckily for you, they make the portal relatively appealing and easy to maneuver. Here, I will briefly introduce and guide you through the application process of OptomCAS.

1. Go to www.optomcas.org and click "Login to the 20XX-20XX OptomCAS Application".
2. Click "Create An Account" and enter your information.
3. Once your account setup is complete, it will prompt you to enter additional information.
4. The first section consists of "Personal Information," which includes "Release Statement," "Biographic Information," "Contact Information," "Citizenship Information," "Race & Ethnicity," and "Other Information." This section should not take much time as long as you know who you are.
5. The following section is the "Academic Information." This section includes "Colleges Attended," "Transcript Entry," and "Standardized Tests."
6. For the "Colleges Attended" section, you will need to enter all the colleges you have attended to obtain a bachelor's degree.
7. The "Transcript Entry" section may take the longest to complete. You will need a copy of your transcript to compute into the system manually. You may use the Professional Transcript Entry service for $69 to automatically enter your transcript, which will save you time and any potential errors. Many students, however, decide to input the information manually. Make sure to read all instructions carefully when entering your transcript.
8. You will then need to order and send an official transcript to OptomCAS, which can be done online through OptomCAS if your university or college institution supports "Credential Solutions," "Parchment," or "National Student Clearinghouse."

If your institution does not help these online transcript services, you will need to mail the transcript directly to OptomCAS (Address: OptomCAS Transcript Processing Center PO Box 9119, Watertown, MA 02471).

9. For the Standardized Test section, enter your exam date information and any other information necessary. You may also be required to submit a TOEFL exam (for non-native English speakers). You may use the GRE, DAT, PCAT, and MCAT as alternatives for the OAT for some schools (refer to the list in Chapter III).

10. The following section is the supporting information, which includes letters of recommendation, experiences, and achievements. Try to get two letters of recommendation from any science professor and two letters from practicing doctors. The requirements for letters of recommendation will be different for each school. Enter all experiences and awards, including any papers you have published.

11. Once you have completed this part, it will prompt you to choose the Optometry Schools that interest you. There is a fee of $180 for the first school you send applications to and then $70 for each additional school. You may choose to add or delete schools later. The application will not be transmitted until the payment is complete.

12. Once you have selected your schools, you will see them appear in the "Program Materials" section. When clicking on each school, you will see the description of the school, questions (which is where you will submit your personal statements), and prerequisites. You will then manually match the courses you previously entered in the transcripts section to the prerequisite courses listed for each school.

13. In the "Home" section for each school, you will see whether that school requires a supplemental application. A supplemental application is an additional application that you must submit through *their school portal*, so keep an eye out for that. These usually cost $40-50 per submission, and they are charged separately from OptomCAS.
14. Once all material is complete, pay the fee, hit submit, and wait! Good job!

Chapter V: What is the OAT?

What is the OAT?

The OAT (Optometry Admission Test) is a standardized examination designed to measure general academic ability and comprehension of scientific information. It is composed of multiple-choice test questions written in the English language and consists of four sections: Survey of the Natural Sciences, Reading Comprehension, Physics, and Quantitative Reasoning. The OAT uses the US customary and metric systems (Imperial System, International System) of measurement.

The OAT has the following sections:

Survey of Natural Sciences (90 minutes): 100 questions
- Biology: 40 questions
- General Chemistry: 30 questions
- Organic Chemistry: 30 questions

Reading Comprehension (60 minutes): 50 questions
- Passage 1: 15~18 questions
- Passage 2: 15~18 questions
- Passage 3: 15~18 questions

Break: 30 minutes
Physics (50 minutes): 40 questions
Quantitative Reasoning (45 minutes): 40 questions

The OAT is scored on a scale of 400 by each section. You will receive scores for all sections, as well as a score for Total Sciences (TS) and Academic Average (AA). The TS score uses only the Survey of Natural Sciences and Physics sections, whereas the AA uses all sections. The Survey of Natural Sciences section consists of Biology, General Chemistry, and Organic Chemistry. Keep in mind – the TS score is by no means the average of the four topics. It is calculated based on a conversion from the raw score.

Applying to take the OAT

1. Visit the ADA OAT website (https://www.ada.org/en/oat)
2. Click "Apply Now" and then "Apply Now" again on the next page.
3. Click where it says, "If you are certain you do not have a PIN, click here" unless you already have a PIN. If you do, log in using your PIN or search for your PIN through the site.
4. Enter your information and submit.

5. You should receive an email regarding your OAT eligibility with your OAT ID and PIN.
6. Search "Prometric OAT" and open the first link, or go to https://www.prometric.com/test-takers/search/oat
7. Click "Schedule" on the left menu bar.
8. The Prometric site will prompt you to input your ADA eligibility ID to schedule a test date. It will not cue you to create an account, so you must do that separately. Making an account will be necessary for taking the ADA practice exam or scheduling a trial run before your exam date.
9. Follow the steps on the Prometric site and schedule an exam date. You can schedule your exam as far out as you would like, so plan accordingly. There are different time slots for each date, so if you are not a morning person, make sure you plan your exam for a later time.
10. I highly recommend you pay for any scores to be sent to schools NOW rather than later. If you already have some schools in mind, make sure you add them to your list of schools. Doing so will help with cost and time in the long run.
11. Once complete, you will receive an email confirmation of your exam date and time. Keep your confirmation code and your eligibility ID saved. You will need these codes multiple times later in the application process.

Chapter VI: OAT Study Tips

OAT Study Resources

There are many study resources that you can use to refresh your memory on basic scientific concepts and study for the OAT. Of course, while the cheapest and most efficient way is to use the materials learned during your undergraduate years, recalling the information and retaining it throughout the years is challenging. So, here are some resources you can use to refresh and relearn the material to prepare you for your big exam better!

Kaplan OAT ($800-$1200)

Kaplan provides one of the most in-depth and detailed preparation materials available online. While the cost may seem extreme, it is worth it for those who are willing to invest. They offer both a self-paced course and an online course. These cover all parts of the OAT while giving ample practice throughout the course. You can choose when to begin, and it comes with a total of 7 full-length practice exams and a "Qbank" with over 1000 questions separate from the exams. Kaplan will send you their "Lesson Book" and "Review Book" at the time of your course registration.

For those who are more conscious about the price, there are options to purchase just the "Qbank" ($99), just the Practice Exams ($149), or the Study Pack that consists of both the "Qbank" and the practice exams ($199). I highly recommend doing all the practice exams – this is where students see the most growth.

Notes about Kaplan: Students who fully engage in the course and focus on studying will see the most improvement in exam scores. After reviewing all the materials, I recommend taking one exam every week and looking over every question, including the ones you got correct. Keep in mind – Kaplan practice exams tend to be much more complex than the actual OAT. Adding 20-40 points on the Kaplan OAT practice exams would be a good representation of how your score will look.

OAT/DAT Bootcamp (Can try for free, upgrade for $497)

The same company runs OAT Bootcamp as DAT Bootcamp. The OAT and the DAT are created by the same test makers (ADA) and have the same difficulty and question types except for a physics section in the OAT and a spatial reasoning section in the DAT. Many students say the practice tests, practice questions, and videos of the OAT Bootcamp are beneficial, as each question has a written and video explanation for better understanding. The questions given in the OAT Bootcamp are like the OAT but slightly harder to ensure you are well prepared for exam day.

Chad's Videos / Khan Academy MCAT Prep

Chad's Videos and Khan Academy MCAT Prep are two excellent and affordable tools to review content. While they do not offer a full-length exam to practice with, their review videos do a great job at explaining concepts thoroughly. Many students go with this option to study and review the material and use other tools to get practice questions in. They offer study sheets and formula sheets for physics and organic chemistry reactions that are extremely helpful, although memorizing all the given equations is unnecessary. These are both fantastic tools to review material and learn concepts that were hard to grasp during your undergraduate education.

OAT Destroyer ($159.95)

OAT Destroyer is an outstanding tool for practicing problems. The OAT Destroyer collection offers practice question booklets with ~400 Organic Chemistry questions, ~450 General Chemistry questions, ~780 Biology questions, ~170 Quantitative Reasoning questions, and ~140 Physics questions. There are also answers on the back of each section for you to review. While this resource provides ample practice problems, it does not offer any review of the material. So, if you forgot all about basic sciences, it could be better to review the material before taking on the OAT Destroyer. Orgoman also has a Facebook page, a fun and interactive tool for you to connect with other OAT takers. They also have a YouTube channel reviewing all concepts, so you can use that to relearn your material! Keep in mind that the OAT Destroyer practice exams are a LOT harder than the actual OAT. It might over-prepare you, so don't be discouraged if you are not getting as many questions correct as you would like. Just keep practicing and practicing, and you will ace the exam!

ADA Practice Exam ($100)

The ADA Practice Exam is a great resource to go over a week or two before your exam date. This practice exam will give you a good idea of what the *actual* OAT questions will look like, and it will increase your confidence if you have taken any of the Kaplan practice exams. Although the formatting of the Kaplan practice exams is most like the actual exam, the ADA Practice Exam truly reflects the difficulty of the actual OAT. Many students find that the concepts in the OAT are rather basic, so the ADA Practice Exam will give you a good idea of what to look over right before your exam. The ADA Practice exam is slightly shorter than the actual OAT.

OAT Study Tips

- Study early! Give yourself two months to study and take practice exams. A month to review material and a month to practice and review is a good schedule to have. You can even start reviewing material earlier if you have the time!
- Register for the exam in May and take the exam in July. Doing so will put you in the earlier part of the application cycle and make you a more competitive candidate. Since most schools are rolling admissions, the earlier you apply and take the OAT, the better!
- TAKE THE PRACTICE EXAMS. Really. The best way to prepare for any exam is to re-enact everything, including the timing. The Kaplan exam format practically mimics the actual OAT format, making you more comfortable during the exam.
- I recommend considering your pacing for the exam sections. Try to use 30 minutes per section in the Survey of Natural Sciences (30 minutes for Biology, 30 minutes for General Chemistry, and 30 minutes for Organic Chemistry). You will find that you might

be better at one subject than another. Use this to adjust your timing based on your comfort and skill in each subject. For example, if you are knowledgeable in Biology but struggle in General Chemistry, use 25 minutes on Biology and the extra 5 minutes on General Chemistry, then have the recommended 30 minutes left for Organic Chemistry. For reading comprehension, use 20 minutes per passage and SKIP any difficult questions. All the answers are in the passage, so use the highlighting function to your advantage (right-click and drag). While you have more than a minute per question on the physics and math sections, SKIP any questions you are struggling with and MOVE ON. The questions are not ordered by difficulty, so spending time on a difficult question will waste your time getting points on easy ones.

- Practice as if you are taking the OAT. Drive to the testing center before you take the exam. Prometric has "30-minute trial runs" where you can go in and walk through the check-in process so that you're not confused come exam day. Wear the same clothes you will be wearing on exam day, and make sure it's comfortable. Take the practice exams at the same time you scheduled your actual exam. Call your testing center and see what kind of scratch paper they give you – for some places, it is only two sheets of paper, and for others, it is a whiteboard. Take your practice exams with the same scratch paper that the testing center provides.
- Arrive early – try to aim for 30 minutes.
- You have a 15-minute tutorial section before the exam. DO NOT SKIP THIS. Use this time to write down all the equations you know on your scratch paper/whiteboard. Write down all the physics equations and the lens/mirror rules, and the

Henderson-Hasselbalch equation. Use this time wisely! Once you are ready, take a deep breath, relax, and get focused.

- The testing center will give you a locker to put your belongings in, except for your ID. Only a thin jacket is allowed. You may not bring in any water, jewelry, or writing utensils. You may bring in earplugs (upon approval by the proctor), but they will offer some there as well. I strongly recommend getting used to taking an exam around other people. There will be other people at the testing center taking different exams at the same time.

- USE THE BATHROOM BEFORE AND DO NOT DRINK TOO MUCH WATER OR COFFEE. The sections before the 30-minute break are the longest. You will need to use the restroom if you go overboard on coffee, and you cannot pause the exam at any time. Use the bathroom towards the middle of the break (around the 15-minute mark) so that you have enough time to get back to your desk on time.

- Go outside during the breaks. Take a deep breath. Relax. Tell yourself you're doing amazing. Ace the exam.

Chapter VII: Personal Statement

"I wrote one for college applications. Must I do it again?!" Unfortunately, yes. But this time, you get to talk about your passions. In my opinion, I enjoyed writing this statement. It was the first time I had ever expressed my love for optometry and my drive to become the best doctor that I could be. So, try to enjoy it! Put in some enthusiasm and channel your passion for optometry into that paper.

Below is the prompt for OptomCAS and some tips and tricks for writing a personal statement. Some schools require the submission of a supplemental personal statement with their supplemental application. Be sure to research your schools, know which ones have a supplemental application, and write your personal statements according to the given prompt. It is usually shorter than the OptomCAS required prompt.

OptomCAS prompt

"Please describe what inspires your decision to become an optometrist, including your preparation for training in this profession, your aptitude and motivation, the basis for your interest in optometry, and your future career goals. Your essay should be limited to 4500 characters."

Some tips and tricks:

- Writing personal statements take time and effort. It is essential to continue improving and revising your essay, but not continuously. Try not to finish writing your personal statement in one day and submit it once you have finished. Write when you're in the "zone," take a long break, come back to it, revise, write some more, edit, take a break, and repeat. This method will help refine your essay without making it sound like an essay you wrote for a midterm exam.
- Use your resources! Your undergraduate or graduate institution should have resources you can use to improve your essay. Often, tutoring services are at no cost.
- Make it into a story. Express yourself – don't just talk about yourself. Tell the admissions team a narrative that they can visualize. They want to see themselves in your shoes and how you became the person you have come to be.
- Don't talk about any negative parts of your application. If you have a bad grade on your transcript, don't mention it. You can explain to the admissions team your situation at the time of your interview.
- Talk about what you learned during your shadowing hours, as well as your passion for

optometry. What kind of doctor do you want to be? What did you see in your shadowing experiences that you wanted to embrace as a doctor?

Sample Personal Statement of a 100% Admitted Student (SUNY, UABSO, UHCO, SCO, PCO)

It was the perfect angle. The soft, golden flare of sunlight peeking through the narrow windows gently brushed her face. Her bright, blue eyes glistened as the warm light complemented the brilliant shades of blue. It was like the ocean, and I couldn't stop staring. As an artist, I had an intense infatuation with the iris. Yet, it wasn't until later that I realized my obsession with eyes. After hours of shadowing and working as a technician at multiple optometry offices, I connected the dots. I realized that my passion for optometry could be traced back to my past experiences as an artist. The last two years of my college career were centered around that passion, gaining new knowledge and experiences as I explored the world of optometry and the depths of the human eye.

My experience started when I visited my optometrist for my routine eye exam. While waiting for my eyes to get dilated, I had this strange epiphany to ask for an internship opportunity at his office. So, I took a deep breath, strapped on my Rollens sunglasses, and took the biggest first step of my life. The result was a mixed blessing. He was not offering any positions at the time but referred me to another office where I had the opportunity to step foot into a new world I was bound to fall in love with.

Since then, my life has been nothing but fascinating experiences and inflating knowledge. After my initial shadowing experience at XXXX, I have been pouring my heart into learning the complex processes of being an optometrist while also gaining wisdom on the administrative side of running a practice. I have been working at a relatively new office, XXXX, as an optician, technician, and lab technician. As the first employee at the practice, I have also performed administrative duties to maintain luxurious patient experiences. As part of my job,

I analyze the inner workings of the practice, learning various skills and acquiring familiarity with every aspect of optometry along the way.

I have also made good use of downtime. While I was not performing pre-exam assessments or assisting patients in the optical, I shadowed the doctors and learned the unique ways they interact with each patient to run their various tests. Their calm tone, as if to avoid inflicting any fear upon the patients and their emphasis on patient education, was truly inspiring – something I would like to do for my patients if the opportunity arrives. While technical skills are essential aspects of the profession, I learned that good communication skills are vital to optometry. Because not everyone is familiar with the process of a comprehensive exam, optometrists and technicians alike need to have the skills to guide patients through each procedure with empathy while fully acknowledging any distress caused by the undergoing tests. Throughout my time working alongside wonderful optometrists, I had the chance to learn different ways to assist each patient so that they can walk out of the office with a brand-new look and a bright smile on their faces.

As an undergraduate student studying neuroscience, I also had the opportunity to take a few courses on the human eye. One of the Neuroscience courses offered at my university, From the Eye to the Brain, taught me both basic and complex knowledge of the ocular components of the human brain and anatomical structures and functions of the eye. This course provided me with a solid foundation of the inner workings of the eyes and reasons for common diagnoses. Since then, I have been applying that knowledge at the practice while my understanding of the field deepens continuously.

My passion for optometry grows endlessly as my goals become more and more apparent. I want to be a reliable doctor – someone my patients can trust for clarity and quality. I want to be an empathic doctor – someone with compassion to provide my patients with pleasant visits. I want to be an intelligent doctor – someone dependable with substantial knowledge of the field. As a student at your optometry school, I aspire to gain expertise to apply those skills and help all patients see the beautiful world that lies ahead of them. I hope to provide every one of my patients with the same level of care that my current practice does, but this time with me as the doctor.

UCBSO Supplemental Personal Statement

"With which one of these core values do you identify most? Accountability - Community - Compassion - Curiosity - Excellence - Humility - Integrity. Provide an example of when that value was challenged and exercised. How will that core value help you persevere through your optometric studies and beyond?"

"Knowledge is Power," my grandfather repetitively told me as he taught me small skills applicable to simply living life. "It brings about efficiency, compassion, and opportunities," he would continue. He never graduated high school – he went straight into the workforce and learned everything on his own, so I never understood his point of view. I wondered how anyone could gain important knowledge without higher education, and his words did not make sense to me. It wasn't until later that I realized that his continued sense of curiosity was what pushed him forward. His eagerness to learn gave him all the knowledge and the power he has today – a value I can now resonate with sincerely.

I had no dying passion for my future throughout high school and the first half of my college career. I wanted to be successful, yes, but I did not have a goal or a plan in mind to get me to that point. At the time, I wanted to keep all my options open and my opportunities broad, which prompted me to participate in various extracurricular activities such as piano, Japanese drums, fine arts, dance, basketball, research, study abroad, and fraternity life. I participated in all these extracurriculars out of curiosity to learn about the plethora of opportunities available to a young, growing child. Yet, with too much interest came indecisiveness. I thought I was simply "good" at everything I participated in and was never the "best." That mentality and the lack of passion for just one field made me a lost child with absolutely no plan in mind, and I simply became a curious kid with a large variety of skill sets.

When I finally found my passion for optometry, I was able to apply my curiosity to maximize my learning about the field. As a local private practice team member, I learned a technician's job while also shadowing the doctors. I learned how to edge a lens and refine my skills as a lab technician while also helping patients directly as an optician. I spoke to many sales representatives and communicated with partnered labs to learn about all the frame and lens designs to be a reliable source of information for our patients. My curiosity allowed me to dive deeper into optometry rapidly, and I quickly gained confidence in my skills as my devotion to the field amplified continuously. I hope to one day apply this sense of eagerness and curiosity to the path of becoming a Doctor of Optometry.

Chapter VIII: Letters of Recommendations

How many letters of recommendation do you need?

Each school has a different number of letters of recommendation required – some have only two, while others require four. Some may only need professor recommendations, while others may require letters from both doctors and professors. After deciding on the schools you will apply to, I suggest you write down how many letters you need from the school that requires the MOST number of letters of recommendation. For example, if school A requires one letter from a professor and one from a doctor, while school B requires two letters from a professor and one letter from a doctor, ignore the requirements for school A and look to fulfill the requirements for school B. That way, while focusing on completing the requirements for just one school, you are also meeting the requirements for the others, thereby reducing stress.

I suggest obtaining two letters of recommendation from professors and two separate letters from doctors. OptomCAS will only allow you to submit up to four letters of recommendation, so by doing this, you should have fulfilled the requirements for all the schools. Also, the more you have, the better your application is! The entire OptomCAS will send your application to the schools you apply to, so getting more letters than what the schools require will give them a better idea of what kind of person you are as a student and future professional.

Honestly, the hardest part of the application was getting letters of recommendation from my university professors. It was difficult for me because I did not catch the attention of many professors until the last semester of the year where there were only about 20 students in my class. I attended a large university, so many of my degree-specific courses had over 100 students for each class. Therefore, getting close to the professors was not an easy task. At the beginning of my college career, I remember when my advisers told me to sit in the front of every class, actively participate and ask questions during lectures, and always stand out to the professors. But to be honest, who wants to do that? Each person has their way of learning and their comfort zone. I, for one, did not like speaking out or getting called on in a lecture hall full of students, and I'm sure some of you feel the same.

For those of you in a smaller college or university, this might not be as difficult. You most likely have smaller class sizes, a better student-to-faculty ratio, and more interaction with peers. Both of my letters from professors were from two of my smallest classes. I got to know my professors personally, so I can see how smaller class sizes are beneficial in this regard. Note that you don't necessarily have to make exceptional grades to get close to a professor. Just make sure you actively learn, understand the material, and prove that you want to learn at the highest level.

All I can say for the doctor's recommendation letters is to shadow or work at a clinic. There are also some opportunities to interview a doctor and talk to them online or in-person, but being present in an office to observe and ask questions would allow the doctors to write a better letter of recommendation for you. Working at a clinic as a technician would be the best, that is, if you are a good employee. The doctors you work with will see your work ethic, professionalism, intelligence, and passion, so I guarantee they will write you an excellent letter for a better application. If you already work at a practice with more than one doctor, try not to get a letter from each of them. Instead, go shadow other doctors (even ophthalmologists), and get a letter from them as well. You want to show that you are professional everywhere and seek knowledge from different types of practices.

How do I ask for these letters of recommendation?

A formal email always does the trick. If you are often with the recommender, ask for a letter in person AND send them an email. Below is a list of items to include in your email:

1. Full name
2. What your relationship is with them
3. What you are applying for

4. What you need
5. How their letter would help your application
6. Resume
7. Curriculum Vitae (optional, or if asked for)
8. The year you attended their class (for professors)
9. The year you shadowed/worked at the clinic (for doctors)

Note that some professors may ask for additional information. They may require you to answer some questions about the course you took from them, such as "what did you feel that you have learned?" or "how did the experience in my class affect your thinking or approaches to your career choices?". They may also require you to send them your personal statement. Whatever they ask for, try to satisfy it. They are only trying to help, and having these additional documents would help them write a great letter of recommendation for you.

Example of an email sent to a professor:

Dr. XXXX,

I hope all is well. My name is Joe Eyeballs, one of your students from "Evolutionary Neurobiology" NEU367V from the Spring 2020 Semester. I first wanted to thank you for sharing your knowledge throughout the semester. I genuinely believe it has allowed me to view the world differently.

I am currently applying to optometry schools across the nation to begin my studies in the Fall of 2021, and the application requires a few Letters of Recommendation. I wondered if you would be willing to provide me and my application with a Letter of Recommendation. I firmly believe that a Letter of Recommendation from you would be a great addition to my application.

Please let me know if you would like to assist me or have any further questions about the application, and I will forward the link for submission.

Thank you!
Sincerely,
Joe Eyeballs
School ID
NEU367V I Spring 2020
Contact information including phone number

Another example on an email sent to a professor:

Dr. XXXX,

My name is Joe Eyeballs, and I am currently applying to five Optometry Schools for the Fall 2021 Semester. I am writing to ask if you would write a letter of recommendation supporting my applications. I believe that the course I took from you in the Fall of 2019, "From the Eye to the Brain," has helped me increase my knowledge in the field of optometry. I have applied concepts from your class to my current employment as an optometric technician and optician. I also firmly believe that you have helped me build a solid foundation for my further studies in optometry.

I have included my resume summarizing my academic and professional experience to help you decide whether to recommend me. I plan on fully submitting my application at the beginning of November.

I genuinely believe that a Letter of Recommendation from you will help my application to my dream profession. Should you decide to recommend me, I will send you any other materials you think would help you in the evaluation process. If you have any questions, please contact me at XXX-XXX-XXXX or via email at joeeyeballs@xxxx.edu.

Thank you for your time and consideration. I look forward to hearing from you!

Sincerely,
Joe Eyeballs
ID

Example of a letter of recommendation request to a doctor:

Dr. Orbit,

I hope all is well and that things are going as smoothly as possible with the County Vision Doc team.

I first wanted to thank you for allowing me to intern, shadow, and learn at your practice back in the Summer of 2018. It opened my eyes to a new world and helped establish a solid foundation for my drive and knowledge of the field of Optometry. Since then, I have been working as an Optician, Technician, and Lab Technician at Dr. Pupil and Dr. Iris's relatively new practice, Restoration Eye Care. As one of their first employees, I was able to apply the knowledge I obtained from you guys while learning the ins and outs of Optometry at a rapid pace.

I am writing to ask if you would write a letter of recommendation supporting my applications. I recently graduated from the University of Texas at Austin with a Neuroscience Major and Business Minor and am currently applying to Optometry Schools for the Fall 2021 Semester. I firmly believe that a recommendation letter from you would help with my application.

Should you decide to recommend me, I will send you any other materials you think would help you in the evaluation process. If you have any questions, please contact me at XXX-XXX-XXXX or via email at joeeyeballs@gmail.com.

Thank you for your time and consideration. I look forward to hearing from you!

Sincerely,
Joe Eyeballs

Once they agree to provide you with a letter of recommendation, you will have to go back into OptomCAS to send a request. OptomCAS will direct you to enter the direct email of the recommender, and they will send them an email with a link to their portal. Here, they will be required to create an OptomCAS account and to submit their letters. Therefore, unless the recommender specifically sends you their letter of recommendation, you will NOT be able to see what they wrote for you. Be sure to choose the right person!

OptomCAS will update you when your account has received a letter of recommendation. Be sure to check your account and your emails often. If your deadline is approaching (whether it be one you set on your own or a school deadline), it is acceptable to remind the recommender professionally. They are most likely busy, so they will make mistakes and be forgetful. If you do so, remind them that you are thankful for their time and ask if they need any other information to help them write a good letter of recommendation. Once you confirm that OptomCAS has received the letters, thank your recommenders, and let them know that you will update them on your application process. They sure would love to hear of any good news and that they played a role in your successes.

Chapter IX: Experiences

Experience is arguably one of the most important factors to take your application to the next level. Most optometry schools review your application based on a holistic approach, so while your GPA and OAT scores may get you through the initial review of your application, your experiences allow you to stand out.

Twenty-five hours of shadowing are usually the minimum requirement for experience. I've seen many acceptances with only the bare minimum, but what would you get out of only 25 hours of shadowing, especially if you have shadowed only one doctor? My advice is: try to shadow as many doctors as you can. Ophthalmologists would provide you with a more surgical approach to optometry and give you a different view of the field. All optometrists practice in their unique way and pace, so you can see how diverse optometry is. Ask questions to every doctor you shadow. It could be as simple as why they chose the field or what made them interested in their mode of practice. They agreed to allow you to learn through them. Abuse the opportunity to ask questions!

You could even shadow doctors unrelated to optometry, such as primary care physicians or orthodontists. During your interview process, the interviewers may ask why you chose optometry instead of other fields of medicine. Observing these doctors may help you solidify your commitment to optometry or open a path to another field.

Again, working at a practice of any kind – private practice, VA, academia, or specialty – is the best form of experience. The possibility of growth while working alongside an optometrist is limitless, and the speed at which you learn is exponential compared to shadowing. The skills you learn as an employee will be helpful in optometry school as well, so you will be one step ahead of your classmates when it comes to clinical application and patient communication.

Optometry schools LOVE extracurricular activities, especially if they involve volunteering. It doesn't necessarily have to be optometry-related – you can do anything that helps the community and makes the world a better place. I was part of a social fraternity, but I had the opportunity to be the Vice President of Philanthropy and Service for a year to impact the community positively. If you do volunteer, try to record your hours actively. While optometry schools do not require you to send proof of volunteer work, you must know what you have done in detail to present it during interviews.

Optometry schools also love research. Many professors are looking for research assistants in their lab, so try to reach out to professors who might interest you and see if they are willing to take you in. Many optometry schools' pillars include "research," so having that as part of your application could be extremely helpful. It would also give you something to talk about during your interviews!

Many schools offer summer/winter programs for undergraduate students to experience optometry first-hand. For example, SUNY hosts EYE-CARE in the summer and CSTEP in the winter, while OSU hosts I-DOC in the summers. These programs act as an introduction to optometry for prospective students and applicants, and your attendance could dramatically boost your application. It would also provide you with an excellent experience for your future as a student.

Chapter X: What are my chances?

It's a tough question. There's no way to quantify it, especially since each school has a different criterium. I can say, though, that if you can get an interview from one optometry school, you can most likely get an interview from all others. Similarly, if you receive an acceptance from one optometry school, you will most likely receive one from the rest. It's hard to generalize the chances of someone getting into an optometry school, but here, I've tried my best to quantify your competitiveness.

This scoring system is not a guarantee of your actual results. Your application is also dependent on your interviews, volunteer hours, grade trends throughout your undergraduate career, volunteer experiences, research experiences, and other factors specific to each school.

Below is the equation I believe is the most reflective of competitiveness:

$$(GPA\ Scale) + (OAT\ Scale) + (Experience\ Scale)$$
$$= Competitiveness$$

GPA Scale calculation:
$$(GPA \times 10) + 10$$

OAT Scale calculation:
$$(OAT\ Score/10) + 10$$

Experience Scale:
0 to 10 hours of experience = 10 points
11-25 hours = 15 points
Over 25 hours = 20 points

70 – Average

This score is an excellent initial goal to have. Try getting more optometric experience! You might get a few offers for interviews, but you could use a boost in your overall applications to stand out more!

80 – Competitive

Keep up the fantastic work! You most likely will be offered many interviews. Continue building your resume and ace your interviews! If you have this score with just the GPA and OAT scores, you're in good shape!

90 – Highly Competitive

Wow! Great job! Research for schools with higher matriculating GPA and OAT averages and rock those interviews!

So, what does this all mean? Nothing. I only have a 70 score. Am I going to have to try again next year? Not necessarily. I have a 90 score! Can I get in for sure? Nope. Remember, optometry school admissions holistically review each applicant, and this is just scoring based on quantifiable statistics. There are many ways you can stand out to each school, so think of what you can do to make your application even better!

Chapter XI: Recommended Application Schedule

Planning is just as important as doing, except sometimes planning is harder to do. So, I went ahead and outlined a schedule that I recommend for you. I recommend at least ten weeks, from start to finish, to go through the application cycle. Just remember that applications open in July, so start applying as early as you can (July-September is early).

Week 1
1. Start looking into schools very casually so that you can schedule a visit with them. Visiting the school will give you the best idea of the environment of the school.
2. Create an OptomCAS account
3. Schedule your OAT.

4. Find the best OAT test prep plan for you. Keep into consideration your daily schedule.
5. Take the Full-Length Kaplan Diagnostic Practice Exam and look over the exam once you are finished. Go through each question, including the ones you got correct. Carefully read through every answer choice and figure out why each choice is correct/incorrect. Doing so will help you learn efficiently. Then, make a note of your strengths and weaknesses to know which subject needs the most focus.

Week 2
1. Complete the "Personal Information" section on OptomCAS.
2. Begin studying for the OAT. Make flashcards. Focus on areas you have trouble remembering

Week 3
1. Complete the "Academic Information" section and the "Supporting Information" section on OptomCAS.
2. Continue studying for the OAT. Make flashcards. Focus on areas you have trouble remembering.

Week 4
1. Look into schools you would like to apply for and begin their application process. Some schools require a supplemental application outside of OptomCAS, so make sure you check the requirements for each school through the OptomCAS portal.
2. Continue studying for the OAT. Make flashcards. Focus on areas you have trouble remembering.

Week 5

1. Begin writing your personal statement. Take your time on this. Write your essay during study breaks or on your break day.
2. Take a Full-Length Kaplan Practice Exam and look over the exam once you are finished. Go through each question, including the ones you got correct. Carefully read through every answer choice and figure out why each choice is correct/incorrect. Doing so will help you learn efficiently.
3. Continue studying for the OAT. Make flashcards. Focus on areas you have trouble remembering.
4. Ask for letters of recommendation. Doing so five weeks before your deadline is early enough for your recommenders not to feel pressured by time.

Week 6
1. Continue writing your personal statement and any other essays that the schools require.
2. Take a Full-Length Kaplan Practice Exam and look over the exam once you are finished. Go through each question, including the ones you got correct. Carefully read through every answer choice and figure out why each choice is correct/incorrect. Doing so will help you learn efficiently.
3. Continue studying for the OAT. Make flashcards. Focus on areas you have trouble remembering.

Week 7
1. At this point, you should be ready to submit your application. ASCO (the OptomCAS company) takes about two weeks to process each application and send it out to each school, so by the time you take the OAT, the schools should have your application on file. You may apply before you complete your OAT.

2. Continue studying for the OAT. Make flashcards. Focus on areas you have trouble remembering.
3. Take a Full-Length Kaplan Practice Exam and look over the exam once you are finished. Go through each question, including the ones you got correct. Carefully read through every answer choice and figure out why each choice is correct/incorrect. Doing so will help you learn efficiently.

Week 8
1. With the application now submitted, all you must do is focus on studying for the OAT.
2. Continue studying for the OAT. Make flashcards. Focus on areas you have trouble remembering.
3. Take a Full-Length Kaplan Practice Exam and look over the exam once you are finished. Go through each question, including the ones you got correct. Carefully read through every answer choice and figure out why each choice is correct/incorrect. Doing so will help you learn efficiently.

Week 9.
1. Continue studying for the OAT. Make flashcards. Focus on areas you have trouble remembering.
2. Take a Full-Length Kaplan Practice Exam and look over the exam once you are finished. Go through each question, including the ones you got correct. Carefully read through every answer choice and figure out why each choice is correct/incorrect. Doing so will help you learn efficiently.
3. Take the ADA Practice Exam and look over the exam once you are finished. Go through each question, including the ones you got correct. Carefully read through every answer choice and figure out why each choice is correct/incorrect. Doing so will help you learn efficiently.

Week 10

1. For your final week, try to take as much time as you can to relax. It's not good to overwork yourself. Look over your flashcards and your notes. You worked hard for nine weeks now, so we just need to get you into the right mindset for the final boss battle. Take a break, hang out with friends, have some fun, but be safe.
2. Schedule a "Practice Run" with your Prometric Testing Center.
3. ACE THAT TEST.
4. You will receive your unofficial scores immediately after you finish taking your exam. Scan the results and email your scores to the admissions offices of your schools of interest. Doing so will get you to the interview process sooner. It takes about two weeks for the official scores to arrive at the schools, and you don't want to be waiting for that!

Here are some scheduling tips that I have for you.

1. Start your application cycle early. OptomCAS applications open in June, so you may start as soon as they open. Remember, the earlier the application, the more competitive you get! The more you wait, the fewer seats are available to you, so it will be harder to be accepted.
2. The information above is just for reference. You do not have to schedule as planned, and it may be hard to do so while you are working or taking classes. You may extend or compress the schedule as you wish.
3. Be confident! Comparing yourself with other applicants might let you down, even though you have competitive stats yourself! Stats aren't everything, and the schools know that too. Keep up your hard work and grind!

4. You can have up to four letters of recommendation. Try to ask your recommenders at least a month in advance. Rushing the letter of recommendation would not give you the best letters. Don't be afraid to provide the professor with a reasonable deadline, and don't be scared to remind them of the deadline if it's coming up.

Once you have completed your application, you're done – with the first half. Now, I have just one piece of advice for you: CHECK YOUR EMAILS, CONSISTENTLY. Schools will contact you only via email, and you will most likely not be using OptomCAS any longer. However, you must remain patient. Some schools may contact you for an interview request within two days, while others may take a month. It is crucial to check your emails daily so that you do not miss any interview scheduling opportunities.

Once you receive an email about scheduling an interview, do so right away. Remember, the sooner you get seen, the more seats will be open for you, and your chances will be better. Try to get your interviews done as early as possible – don't wait!

Chapter XII: Interviews

You've made it this far. Congratulations! Quick statistics for you: only about 30% of all applicants make it past the initial reviewing process, so be proud of yourself. They know who you are on paper, and they know your academic abilities. Now, you must show them your personality. Most of the time, your interviewers are doctors of that school, so they understand your position and recognize the nervousness and enthusiasm you are feeling.

To be honest, these interviews were, without a doubt, one of the most exciting interviews I have ever experienced. The admissions team is entertaining, and the schools are usually very welcoming. They want you there as much as you want to be there, so have fun with it! Remember, they selected you to interview for a reason, and now they just want to get to know you.

Optometry is a very social field. You can't become an optometrist without having any "people skills." The admissions team is essentially looking for these "people skills" in applicants. Is the applicant professional? Are they outgoing? Are they able to hold a conversation? Are they able to present themselves in a well-mannered, positive way? Are they showing interest in our school? Are they who they said they are in their personal statement?

Keep in mind that the school already knows everything about you – your undergraduate school, GPA, OAT scores, extracurriculars, and interests. What they don't know is what you haven't told them yet. What kind of person are you? What are your strengths and weaknesses? Why didn't you choose a different medical path other than optometry? Your interviewer will most likely ask you these types of questions during your interview, so try to brainstorm your answers to questions that your resume or transcript does not reflect.

You can almost treat this interview as a conversation with guided questions, which means you can ask questions, too! I would recommend you ask questions throughout the interview and at the end, whether it be about the interviewer or the school. However, make sure not to ask anything with answers on the school website – this just emphasizes that you did not research the school before the interview. You may ask for clarification about special programs they offer but try to be informed about the school as much as you can beforehand. While this entire process might rustle your nerves, remember that this interview is very casual. During one of my interviews, for example, I had a 10-minute conversation about my hometown. During another one of my interviews, I presented my glasses collection and told them about each frame and specialized lenses that I had put in them. It was almost like a lunch-with-doctors event. We all had fun!

Some schools will ask you if you are applying to other schools. Be honest – they don't care as much as you think they do. If anything, applying to two to four other schools is expected and would show them that you are passionate about entering the field and are actively trying to become an optometry school student. With that said: it is vital to present to them your reasons for applying to the school and why your acceptance there would benefit you.

Example interview questions from the interviewer:

1. Tell me about yourself.
2. What has been the most challenging experience in your life?
3. Why optometry instead of ophthalmology?
4. Why this school?
5. What was the most interesting experience you had with a patient while working as a technician?
6. Tell me about your research.
7. How do you cope with school workload and stress?
8. Tell me about your leadership experiences and how they have helped you become who you are.
9. How do you study and prepare for exams?
10. What was your most valuable course in prior education?
11. Who has influenced you the most and why?
12. What can you provide to our school?
13. Why do you think you are a good candidate for our school?
14. What are your greatest strengths and weaknesses?
15. What are you looking for in an optometry school?
16. What other optometry schools are you applying to?
17. Your grades seem to be low during XXX semesters. Can you explain what happened and what you have learned from it?
18. Do you have any other piece of information you would like me to know about you?
19. Do you have any questions for us?

Example questions to ask the interviewer:
1. Why did you choose optometry?
2. Did you choose a specialty, and why?
3. How long have you been working at _____ school of optometry?
4. What do you think is the best part of being in that city?
5. What do you think is the best part of being a member of _____ school of optometry?
6. What is the most interesting/funniest patient case that you have experienced?
7. Is there anything else on my application that I can clarify for you?

These questions are simply example questions and are not necessarily all the questions that they could ask. They cater each question towards each applicant based on grades, experiences, and personal statements. You may also ask the interviewers some other questions based on the flow of conversation or if you think of any additional questions. You can even ask them to show off your glasses collection as I did! They love that kind of stuff!

Chapter XIII: Tips from Past Optometry School Applicants

3.56 GPA | 340TS | 330AA | PUCO, AZCOPT

Experiences: 20 hours shadowed at an optometric dry eye clinic, informational interview from another optometrist
Extracurriculars: Pharmacy technician for 2+ years, Pre-Pharmacy Society Club member, MS (Multiple Sclerosis) Exercise Clinic volunteer, Oregon Museum of Science and Industry volunteer.
Applied: PUCO, UCB, OSU, AZCOPT
Accepted: PUCO, AZCOPT

Advice and tips: "Apply early, do your research, and know-how to sell yourself during interviews."

3.6 GPA | 330TS | 350 AA | ICO, SCCO, SUNY

Experiences: 100 hours interning at an optometric office, hospital volunteering
Extracurriculars: Youth group leader, mission trips, chemical biology research, published a peer-reviewed paper
Applied: ICO, SCCO, SUNY
Accepted: ICO, SCCO, SUNY
Advice and tips: "Plan ahead (emailing LOR requests, filling out OptomCAS experiences/transcripts, studying for OAT) and get as many different optometric experiences as you can!"

3.6 GPA | 340TS | 330 AA | OSU, NOVA, UHCO

Experiences: Two research publications in Jama Ophthalmology
Extracurriculars: Dance team
Applied: OSU, NOVA, UHCO
Accepted: OSU, NOVA, UHCO
Advice and tips: "Practice interview questions!"

3.87 GPA | 370TS | 370AA | UHCO, ICO, OSU

Experiences: Optometric Technician (1.5 years), shadowed (~20 hours)
Extracurriculars: Vice President External of sorority, POPS
Applied: OSU, UHCO, ICO
Accepted: OSU, UHCO, ICO
Advice and tips: Believe in yourself! As cheesy as it sounds, half the battle of applying is believing that you have a shot. You do! If you truly have a passion and want it, you will find a way to make it there. No application is perfect, and everyone has their weaknesses. Don't let it discourage you; I appreciate how schools look at your application as a whole. Try to be a competitive applicant in as many aspects as you can. Truly be yourself in interviews; they're trying to impress you just as much as you are trying to impress them. If you are fortunate enough to have different schools to choose from, know that it's not always an easy decision and there may not be a perfect choice, but there will always be the better one. Decide what your priorities are and choose the school where you feel you will excel most. Optometry school is really what you make of it, and it can be some of the best four years of your life if you're in an environment to thrive. Also, ya know the usual… Apply as early as you can. Y'all got it! Good luck future colleagues!

3.8 GPA | 370TS | 370AA | UHCO, UIWRSO

Experiences: Optometric technician for 1 year
Extracurriculars: None
Applied: UHCO, UIWRSO
Accepted: UHCO, UIWRSO

Advice and tips: Try your best but don't overthink too much!

3.74 GPA | 370TS | 350AA | UHCO

Experiences: None
Extracurriculars: None
Applied: UHCO
Accepted: UHCO
Advice and tips: Be sure you're really committed to optometry. It's definitely not easy.

3.14 GPA | 320TS | 330 AA | ICO, UIWRSO, NSUOCO

Experiences: Worked as a medical assistant and CL manager for an optometrist for 2+ years
Extracurriculars: None
Applied: ICO, UIWRSO, NSUOCO
Accepted: ICO, UIWRSO, NSUOCO
Advice and tips: Own up to any mistakes you made in college, such as declines in your GPA. Schools will most likely ask you what the cause was, don't give excuses. Just tell them the reason and how you learned and improved from that situation. Also, for interviews, just be yourself, but remember to stay professional. Schools just want to get to know you better, and it's a very welcoming environment

3.82 GPA | 350TS | 330 AA | NECO, WesternU

Experiences: 30 hours of shadowing
Extracurriculars: None

Applied: NECO, UMSL
Accepted: NECO, UMSL
Advice and tips: Apply earlier <3

3.90 GPA | 310TS | 310 AA | UHCO, SCO

Experiences: Shadowed an OD for one summer and worked as a tech for 1 year
Extracurriculars: None
Applied: UHCO, SCO
Accepted: UHCO, SCO
Advice and tips: Don't stress over the OAT! It matters, of course, but you can still be accepted with just an "okay" score

3.55 GPA | 340TS | 350AA | UIWRSO, UHCO, SCO, IUSO

Experiences: 1 summer of shadowing, hospital volunteering
Extracurriculars: None
Applied: UIWRSO, UHCO, SCO, IUSO
Accepted: UIWRSO, UHCO, SCO, IUSO
Advice and tips: Don't worry if you don't have any experience

3.7 GPA | 320TS | 320AA | SCCO

Experiences: Worked for an optometrist
Extracurriculars: None
Applied: SCCO, AZCOPT, ICO, PUCO, SUNY
Accepted: SCCO
Advice and tips: Go for it!

4.0 GPA | 360 TS | 350 AA | UMSL

Experiences: Target Optical retail optician (522 hours), private practice optometric technician (630 hours), shadowing at a private practice (18 hours)
Extracurriculars: Pre-Optometry Society President, Service Fraternity Member, Youth Group Leader, Student Government Leader
Applied: UMSL
Accepted: UMSL
Advice and tips: Apply as early as you can, but don't rush through the application! I applied/took my OAT in late September/early October, and our class had barely a fourth of the class admitted. Keep track of the hours you work, shadow, and do extracurricular activities as you go because that took a lot of time to calculate on my OptomCAS application. Try to ask for letters of recommendation and fill out transcript information right when the application cycle opens to get that portion out of the way, even if you haven't taken your OAT yet! I submitted my entire application two weeks before taking my OAT and got offered an interview the same day I emailed UMSL my scores. Relax, you will do great!

3.36 GPA | 350TS | 350AA | PUCO, AZCOPT, NECO

Experiences: Optometric technician, other work experiences, varsity athlete
Extracurriculars: None
Applied: PUCO, AZCOPT, NECO, UHCO waitlist
Accepted: PUCO
Advice and tips: Apply early! You can submit one application at a time, so you can prioritize where you apply to save time and money. Do your research a few years

before applying to optometry school, so you know what you like. Study with Bootcamp OAT and Chad's Videos/free quizzes. Get in touch with admissions from as many schools as soon as you can (even the ones you aren't interested in). You'll get spam emails, but occasionally, they'll have virtual seminars that are saturated with application tips. Show your passion and knowledge for optometry in your personal statement and interviews!

3.95 GPA | 380TS | 400AA | SCO

Experiences: Part-time optometric technician for 1.5 years at a Costco practice, shadowed for ~60 hours at an ophthalmology, private, and corporate practice

Extracurriculars: Vice-president of pre-optometry society, member of a band started with friends that performed ~20 times a year, volunteered with local park clean-ups and for a local pet sanctuary, worked as a STEM after-school program instructor

Applied: SCO

Accepted: SCO

Advice and tips: Don't overstress and apply early! I was pretty anxious about each step of the application process which led to me pushing things back. However, the earlier you apply, the better your chances and the more likely you are to receive scholarships. I would also say to put less emphasis on your stats and extracurriculars and more on your personal statement and interviews. As long as you have around average stats and experiences, you should step worrying about "am I competitive enough?". I found that what most admissions committees were looking for was a good personality, a real love for optometry, and qualities that would make you a great doctor and practitioner. Show them that in your personal statement and interview! For the OAT, I would really recommend

OAT Bootcamp + Chad's videos for extra physics help. Bootcamp gave me everything I needed to succeed and no more. The practice tests, practice questions, and content videos are extremely helpful. I love that each question has a written and video explanation. I also found the questions to be similar but slightly harder than the OAT so that you are just slightly over-prepared, and the real test feels like a breeze.

3.82 GPA | 300TS | 310AA | OSU

Experiences: 21 hours of shadowing with five different optometrists, I-DOC program, and GPS Day at OSU
Extracurriculars: 4 years as a golf student-athlete, POPS secretary, student-athlete advisory committee member, ~50 hours volunteering with various programs
Applied: OSU
Accepted: OSU
Advice and tips: I would definitely say don't get discouraged by comparing yourself to other applicants! Everyone has their own path and is a competitive applicant in their own way! Schools are looking for diversity and want people that have their unique stories and backgrounds. Make sure you elaborate on yours in your personal statement and interviews! Regarding the OAT, practice, practice, practice. I wish I had done that sooner in my OAT studying schedule. You can do it, guys!!!

3.1 GPA | 310TS | 320AA | PCO, SCO, ICO, NOVA, NECO

Experiences: 2.5 years working as an optometric technician and optician, ~20 hours shadowing

Extracurriculars: Secretary/Treasurer of Pre-Optometry Club for three years, Undergraduate Activities Board three or three years, volunteer for Habitat for Humanity, Director of a leadership and diversity program

Applied: PCO, SCO, ICO, NOVA, NECO

Accepted: ICO

Advice and tips: Apply early!!! Also, write a killer personal statement – include a story, make it personal to you, and emphasize why optometry and what you want to do with the profession. I had my essay edited by almost 20 different people, and each time someone found a grammar mistake or commented on how something could be written better. Don't be afraid to ask for help and advice from your friends and peers. Reach out to the admissions offices of schools you are interested in, most of them will be willing to help you and tell you where your application has its weaknesses and things you can do to make yourself a better applicant. It also helps to familiarize your name to them, which can be to your benefit! When it comes to interviews, practice, practice, practice!! Make a list of possible interview questions and your responses to them, and research the schools and what makes you interested in them. Good luck!

3.2 GPA | 390TS | 380AA | SUNY, UABSO, SCO, PCO, UHCO

Experiences: 80 hours of shadowing two different optometric practices, 1.5 years working as an optician, technician, lab technician, and assistant manager, researched at a university in Japan for four summers since sophomore year of high school, published a peer-reviewed paper in 2020

Extracurriculars: President and Vice President of Service of a fraternity, Society of Asian Scientists and Engineers member

Applied: SUNY, UABSO, SCO, PCO, UHCO
Accepted: SUNY, UABSO, SCO, PCO, UHCO
Advice and tips: First off, apply early. The earlier you apply, the more competitive you get and the more chances you get for scholarship opportunities. Find what you really value in each school – for me, it was the clinical experience and tuition. SUNY, SCO, and PCO offered great clinical exposure, while UABSO was attractive to me because of their prestigious research opportunities. My GPA was not that great coming out of undergrad, but I proved myself by studying hard for the OAT and ended up getting the honors scholarship for SCO and acceptance into SUNY (which usually has a high GPA class profile). The numbers can balance each other out. Having experience in the field will help significantly with interviews and the overall application, so try your best to get as much shadowing or work experience as you can. As for the interviews, just be yourself! All the interviews I had were super chill, and they just really want to get to know you. I even showed off my glasses collection and explained the different types of lenses I had in every single one! If you ever visit a college, make sure you talk and ask questions. They tell you that they don't pay attention to how you interact with everyone… They're lying to you. Be interactive and show enthusiasm to be there. Another thing – don't stress about a few bad grades on your transcript. Don't blame your professor for your poor grades, explain what led up to that grade, and own up to it. Then present your growth and show that you learned from that experience and can do well in optometry school. Keep pushing and less stressing! Good luck!

Chapter XIV: I got in, now what?

First, let's celebrate. Congratulations! Next, let's decide on which school you would like to attend. If you have one school you are already particular about, you're done! Pay your deposit, enjoy a glass of wine, and relax until orientation. If you are unsure, you're almost there! The most crucial part of choosing an optometry school to matriculate into is to list out, in order, what is important to you. Tuition, location, clinical experience, class size, city – each of you probably has something you are looking for, so put that at the top of your list. If you have no preference, make a table with the pros and cons of each school to see what stands out most to you. Remember, these four years are your four years, and optometry school is what you make out of it. Think with your brain and go with your heart.

How to survive optometry school

Congratulations! You made it! Now you're just four years away from becoming a Doctor of Optometry. But fear not, it won't be easy. Optometry school is your busy week during undergraduate studies, except all the time. I get to study what I want to, right? *Most of the time.* Do I get to take a break? *Sometimes.* Is it survivable? Most Definitely.

Optometry school is medical school, no matter what other people say. It's a full-time job with no overtime pay. There's always something you can do, and procrastination is not an option. If you thought you knew how to study in undergrad, you're probably wrong. The material is dense, and you need to retain all information presented to you for the rest of your career.

I make it sound like hell. It's tough. But you must keep in mind that you're not alone. Thousands of other students are going through optometry school with you, and your classmates will be there to experience the rigor of school alongside you. You're passionate about what you're learning. You love the eyes. So, it will be okay!

I think becoming an optometry school student will better explain the experience of being an OD student. Sure, I can give you advice on succeeding in optometry school, but every school is different. Some schools start clinical integration as early as their first year, while others might not start it until their second year. Some schools might teach gross anatomy before neuroanatomy, while other schools might blend the two courses. Each school has its unique teaching method, so you must adapt accordingly and learn actively.

I have one last piece of advice for you. Enjoy it! You only get to experience optometry school once and do it with the same people for four years. Make use of your time in a new environment and possibly in a new city. Get out of your comfort zone and try something new. Use the four years to grow and mature. Four years is a long time to grow but also a short time to experience. See you on the other side!

References

"Profiles of Applicants." *ASCO,*
    https://optometriceducation.org/future-
    students/resources/profiles-of-applicants/.

Made in the USA
Middletown, DE
28 February 2023